YOUR KNOWLEDGE HAS VALUE

- We will publish your bachelor's and master's thesis, essays and papers

- Your own eBook and book -
  sold worldwide in all relevant shops

- Earn money with each sale

Upload your text at www.GRIN.com
and publish for free

Bibliographic information published by the German National Library:

The German National Library lists this publication in the National Bibliography; detailed bibliographic data are available on the Internet at http://dnb.dnb.de .

Imprint:

Copyright © 2017 GRIN Verlag, Open Publishing GmbH
Print and binding: Books on Demand GmbH, Norderstedt Germany
ISBN: 9783668603448

This book at GRIN:

https://www.grin.com/document/384572

Sharyar Ahmed

Integration of Thermal Energy Storage into Energy Network

GRIN Publishing

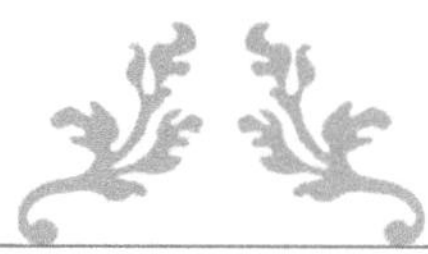

INTEGRATION OF THERMAL ENERGY STORAGE INTO ENERGY NETWORK

Research Report

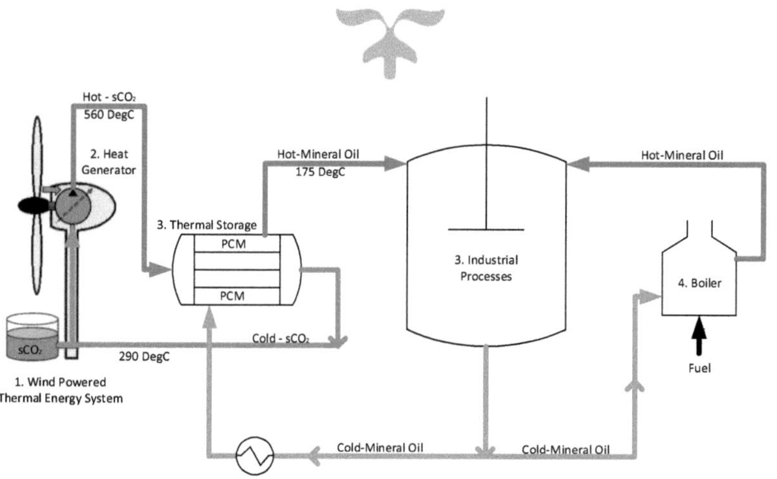

AUGUST 17, 2017

SHARYAR AHMED

University of Birmingham

Abstract

With the increasing world population growth, energy supply can face many challenges in future regarding a sustainable dispatch and in mitigating climate change.

Fossil fuel plays a major role in the society, and they are commonly used for the heat and power production for the commercial and residential applications. According to UKERC Research Report, heat accounts for around 47% of the total energy consumption in the UK and 33% of carbon emissions. Approximately, 80% of this heat account for space and water heating. At the same time, a significant amount of heat is being used by the industry sector 24%, for industrial processes. The corporation of thermal energy storage system is very beneficial for CO_2 emission reduction and energy saving. And, it is the key technology to balancing the energy system stable with the increasing capacity from wind.

This research report uses the concept of Wind-Powered Thermal Energy System (WTES) recently proposed by (2015, Okazaki et al) with Latent-Thermal Energy Storage (TES) design to produce stable heat for industrial application and also, looked at the economics of h-WTES (Direct heat production) and electrical-WTES (Indirect heat Production using electrical heater) with thermal energy storage for heat generation and compared the results with the traditional gas boiler.

This paper demonstrates the potential and probable costs of different WTES with thermal energy storage and a design of tube-in-tank phase change thermal storage system by using a previously developed an effectiveness-NTU method potentially applicable for the industrial sector.

From the results, it was identified that it is more economically feasible to generate heat from h-WTES (Direct heat production) with thermal storage than e-WTES (indirect heat production). 56.2 £/MWh-t and 63.36 £/MWh-t, the levelized cost of heat for h-WTES and e-WTES respectively. Also, the result suggested that the renewable technologies with TES present the most expensive LCOE for heat production compared to the industrial gas boiler which has the LCOE of 15-20 £/MWh.

The result from the second part of the research shows that the technique effectiveness-NTU can accurately predict the average effectiveness of the system (performance) with high heat transfer rate for discharging and charging processes (115kW). Therefore, the design can effectively be integrated into WTES to provide a significant amount of heat demand for industrial applications. It was also, found that latent-TES can hold up to 6.5 times more heat than sensible-TES. Finally, it was identified that by having a 2MW-WTES with latent-TES for the food industry, can reduce 625 tons of CO_2 emission per year from burning fuel such as natural gas.

Table of Contents

List of Figures

List of Tables

1.0 Introduction

1.1 Research overview

Fossil fuel plays an important role in the society, and they are commonly used for the heat and power production for the commercial and residential applications (Li and Zheng, 2016). According to UKERC Research Report, heat accounts for around 47% of the total energy consumption in the UK and 33% of carbon emissions. Approximately, 80% of this heat account for space and water heating. At the same time, a large amount of heat is being used by the industry sector 24%, for industrial processes (Eames et al., 2014). To achieve the UK target of 80% reduction in carbon emission by 2050, a significant drive towards decarbonise heat production is required. The Scottish Government have introduced a 'Heat Hierarchy' to meet the emission target which includes: reducing heat demand, supplying heat efficiently at least costs to consumers and using low carbon/renewable sources (Radcliffe, 2015).

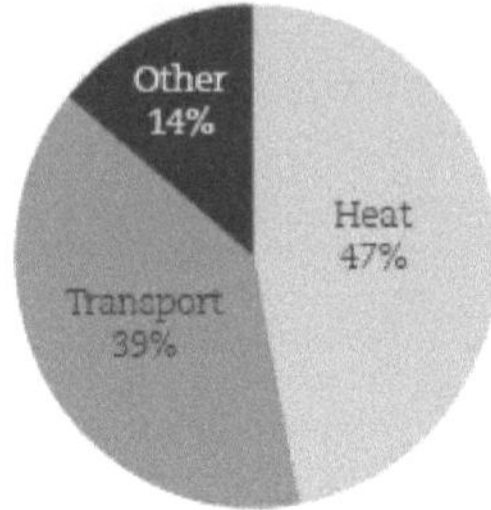

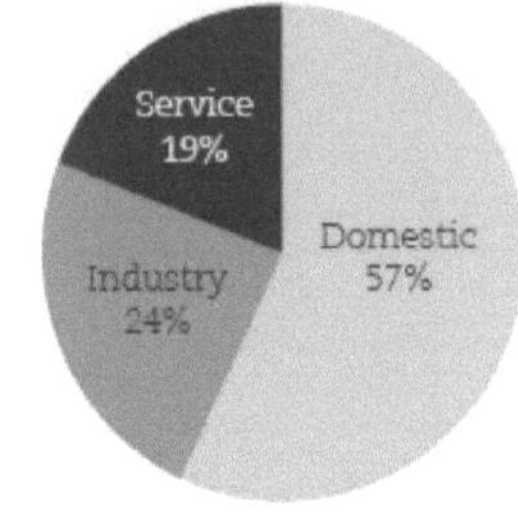

Figure 2: Energy consumption by End Use, 2015. (Eames et al., 2014)

Figure 1: Heat Use by Sector, 2015, UK. (Eames et al., 2014)

In recent years, there has been a significant drive towards renewable technologies such as solar, wind power, biomass, hydro and also nuclear energy. These can lower the carbon emission and allow UK to achieve their European target of 20% energy from the renewable by 2020 (EU, 2008). However, most of the renewable energy sources are intermittent and can't be used as the baseload energy source because of this, thermal energy storage (TES) has become a crucial technology for energy crisis and, playing an important role for renewable energy applications as it makes the operation more efficient (Tian and Zhao, 2011). In brief, TES system stocks the thermal energy by cooling or heating a storage substance for later usage.

Integration of TES with renewable sources can balance the difference between the global energy demand and supply in various applications. This has attracted the industry and scientific community, and it is seen as one of the innovative solutions for energy management. In general, TES is the warehouse between production and consumption of energy and this allows. (Radcliffe and Li, 2015)

- Energy which is generated by renewables at times when demand is low, to be provided as heat or electricity at peak times.

<u>**Thermal Energy Storage (TES) Technologies**</u>

TES system stocks the thermal energy by cooling or heating a storage substance so that the energy can be used later for power generation or heating and cooling applications. TES can be classified as sensible heat storage, latent heat storage and thermochemical storage. (IRENA, 2013)

<u>Sensible Heat Storage System</u>
In this system, the energy is stored/removed by lowering/raising the storage medium temperature in a solid or liquid with no phase change or chemical reaction taking place. The quantity of energy stored in the medium depends upon the amount of the storage substance (m), specific heat capacity of the material (C_p) and the temperature difference (dT) (Liu et al., 2016). A substance with high C_p can store a large amount of heat for a given temperature change. Storage materials include water, granite, concrete, oil, molten salt, rock, etc. (Eames et al., 2014). The main advantages of this system are its simplicity and low cost (cost for sensible heat storage ranges between 0.1-10 €/kWh). Disadvantages are low energy density (3-5 times less than PCM and require large volumes) and heat losses.

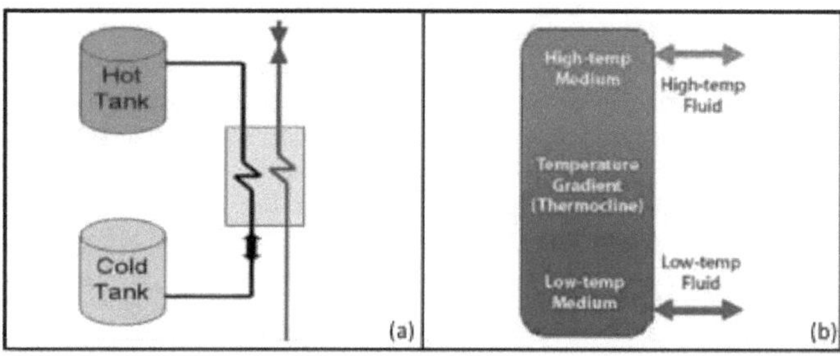

Figure 3: tank storage system. (Herrmann, Kelly and Price, 2004)

<u>Latent Heat Storage Systems</u>
Latent heat storage systems, store and release thermal energy through the state change process of the material which is called phase change materials (PCM). PCMs have an advantage over the sensible heat storage materials as they have high energy storage density at a constant temperature (Liu and Rao, 2017). Due to the phase change, the PCM is typically different from HTF, where PCMs are encapsulated in containers with HTF flowing over them or by using a heat exchanger, inserted into a store full of PCM materials (Eames et al., 2014). The advantages of PCM are larger energy densities, higher efficiencies and constant discharging temperature. The disadvantages of PCM are low thermal conductivity and requires a large heat exchange area. Other drawbacks for PCMs are expensive and corrosive (Sharma et al., 2009). These can be designed for long term energy storage, applicable for industrial application. Storage cost range between 10-50 €/kWh.

Phase transformation	Latent heat	Volume change	Advantages/disadvantages
Solid-solid	Small	Very small	Only crystalline state changes, no liquid flow
Solid-liquid	Large	Small	Flow in liquid state
Liquid-gas	Very large	Very large	Large volume in gaseous

Table 1: Phase transformation material comparison.

<u>Thermochemical storage</u>

This system stores thermal energy by using thermochemical reactions (e.g. adsorption or the adhesion of a substance to the surface of another solid or liquid) and offers higher storage capacities than PCMs, operation temperatures of more than 300°C and efficiencies from 75% to 100% (NICLAS NITTO, 2017). A current review article on thermochemical storage (TCS) (Prieto et al., 2016) shows that the system is still very much at the laboratory level and none of them seems to have the characteristics to become the storage system.

A full review/summary of thermal energy storage with the various application is present in the appendix.

1.2 Research Aims and Objectives

The purpose of this research project is to use the idea of the Wind Powered thermal energy system (WTES) proposed by (Matsuo and Okazaki, 2017) for generating electricity, to directly use the thermal energy (generated) to supply heat demand in the industrial sectors such as food and paper industries which require heating of a fluid stream (hot water, hot air streams) and heating of some reservoir. Currently, heating systems in the industries are based on heating of a fluid/air streams from a boiler, which uses a significant amount of fossil fuels like coal, oil and gas. According to (Eames et al., 2014), UK is considered to be the windiest country in the Europe and could power itself several times over using wind and a 2MW$_{th}$ of wind power energy system is considered to be enough to supply the average heat demand of an industrial plant (Gov.uk, 2017). The limiting factor for the wind energy is the non-continuous supply of energy. Therefore, accurate design and sizing of the thermal storage system are required, taking into consideration the particular demand profile.

Overall, there are various studies which have investigated the integration of Thermal Energy Storage (TES) with technologies such as CSP, CHP, wind power, etc., for the generation of electricity. However, there has been an insufficient investigation on TES integration with renewable technologies to provide decarbonised heat for industries. Taking the idea from (Matsuo and Okazaki, 2017) and (Okazaki, Shirai and Nakamura, 2015) this research will investigate the integration of PCM-Thermal Energy Storage (high energy storage density at a constant temperature) into Wind-Powered Thermal Energy System (WTES) for industrial application to provide decarbonised heat. Which has not been discussed by the author yet. This research consists of two different parts:

1. The potential and probable costs of different Wind thermal energy system with thermal energy storage and comparing it with the traditional heating source to provide useful heat.
2. The designing of the latent-TES using the effectiveness-NTU modelling suitable for industrial application. And, comparing the results with the sensible -TES.

1.3 Project Description

<u>Overview of WTES with Thermal storage</u>

A WTES with thermal storage system can convert wind power efficiently into thermal energy which could be stored for low cost and stable heat generation. Integration of WTES with thermal storage into industrial application consists of low-cost heat generator employing inductive heating technology (electromagnetic induction), a PCM thermal storage system for stabilisation and a boiler for further heating. According to (Matsuo and Okazaki, 2017), this is an economical (the cost is 10% lower) solution than the combination of thermal plant and wind power.

The novel configuration of the process is shown in the figure below Fig 4. The heat generator employing inductive heating technology converts the rotating energy to the thermal energy directly at the top of the tower without the use of fossil fuel. The heat energy produced by WTES is indirectly stored in the PCM storage system with sCO_2 (20MPa) as an HTF, because sCO_2 is chemically stable, low-cost, reliable and can work under high temperature making it a desirable candidate. Further, the stored heat is transferred in the Cold-Mineral Oil (As it provides reliable heat transfer and has the performance benefits of low fouling, high thermal stability, practically non-toxic and environmentally friendly) coming from the industrial process as an HTF which is used for drying, washing, sterilising, bleaching, etc. The fossil fuel boiler is used when the plants heating demand is not meet. The minimum inlet temperature of 140^0C was considered for mineral-oil, and the design inlet temperature into the industrial process was defined as 175^0C, which is the average temperature of the food and paper industry.

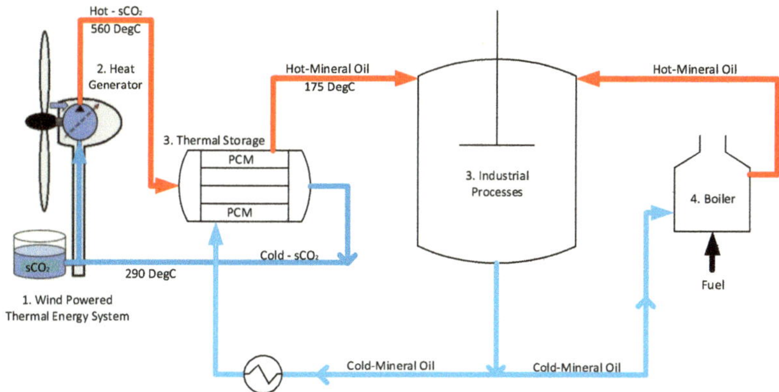

Figure 4: The basic configuration of Wind Powered Thermal Energy System (WTES) with PCM-Thermal storage for industrial processes

1.4 Review of the literature

Overall, there are various studies which have investigated the integration of Thermal Energy Storage (TES) with technologies such as CSP, CHP, wind power, etc., for the generation of electricity (IEA-ETSAP and IRENA, 2015) (Liu et al., 2016) (Li and Zheng, 2016). However, there are not many studies on TES integration with Wind-Powered Thermal Energy system (WTES) which is recently proposed by Okazaki et al. (directly converts wind energy to heat), and their economic and technical feasibility is also unknown compared with other conventional wind systems.

Direct conversion of the wind to heat using a Joule Machine was developed by (Chakirov and Vagapov, 2011) which converts the wind energy directly into cost-effective thermal energy using heat generator driven by a wind turbine for domestic users. Further, (Meibom et al., 2007) this study analysed the economic value of using heat pumps and electrical heat boiler integrated into wind power for heat production. Moreover, (Hedegaard and Münster, 2013) showed the significance of integrating individual heat pump and hot water tank as storage system into wind power. There are also few patents presents on the concept of WTES. (Lee; Jean L, 2012), patented a novel wind turbine WTES concept to produce electricity from heat. Recently, (Matsuo and Okazaki, 2017) (Okazaki, Shirai and Nakamura, 2015) proposed a WTES system with sensible heat storage (not in detail) which directly converts wind energy into heat for electricity generation.

Additionally, (Colella, Sciacovelli and Verda, 2012) (Liu et al., 2015) (Belusko et al., 2016) these studies show the investigation/design of tube-in-tank latent thermal storage for CSP, solar tower plants, CHP and district heating applications by the use of effectiveness-NTU numerical modelling technique.

All of these studies have shown that the concept of WTES with latent thermal energy storage to produce heat for industrial applications has not been studied yet and also the economics of WTES and e-WTES (Wind Turbine with electrical Heater) with thermal energy storage for heat generation is unknown.

2.0 Levelised Cost of Energy

Wind thermal energy system (WTES) with thermal energy storage for heat generation has not yet been studied in depth, and their economic and technical feasibility is unknown compared with other conventional wind systems. Since there is very little research on economic benefits, this part of the thesis will include the potential and probable costs of different Wind thermal energy system with thermal energy storage and comparing it with the traditional heating source to provide useful heat at 175^{0}C.

The heat can be produced in two different ways indirectly or directly. The indirect generation typically requires an electrical generator to convert rotating energy into electrical power and utilises this to generate heat Fig 6. On the other hand, direct generation produces thermal energy without any intermediate like electric generators Fig 5. The full configuration of both systems is shown below in the diagram.

Direct Heat Production (WTES)

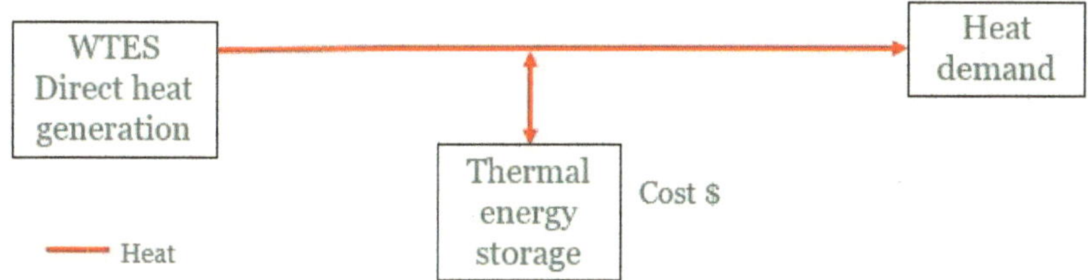

Figure 5: Heat wind turbine with thermal energy storage.

Indirect Heat Production

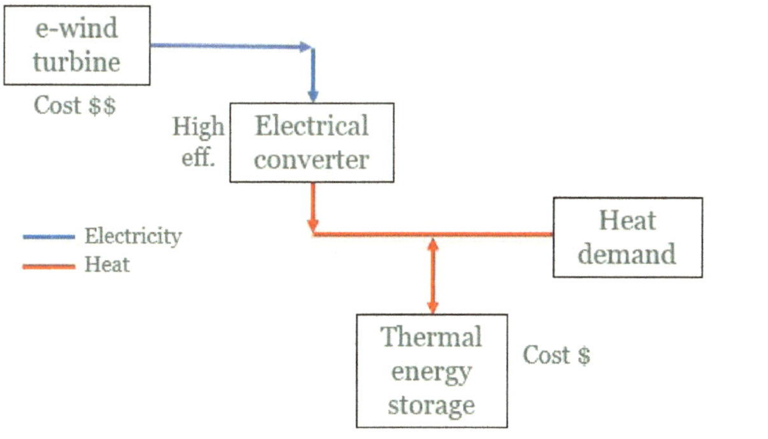

Figure 6: Electrical Wind Turbine with thermal energy storage to meet heat demand.

2.1 Methodology

This section describes the economical parameter which has been used to fulfil the economic assessment of two different WTES concepts. The methodology which has been used to assess these technologies are the Levelized Cost of Energy (LCOE). This method allows different renewable energy technologies to be compared and it is widely used for policy and modelling development.

Levelized Cost of Energy

The LCOE method which has been used in this research follows the procedure described by (gov.uk, 2016). LCOE for a certain technology is the ratio of the total costs of that particular technology which includes the operating and capital costs, to the total amount of energy which is expected to be produced over the lifetime of the plant. Expressed in net present value term and converted into an equivalent unit of cost of generation in £/MWh (gov.uk, 2016). The main benefit of this method is its transparency and simplicity.

This approach does not consider wider costs such as network investment, the full cost of system balancing. And also, does not recognise the revenues from the sale of energy or other sources. The formula and the assumptions which are used for calculating the levelized cost of energy is described below.

The capital costs include: Pre-development costs, Construction costs. The operating expenses include: Fixed opex, Variable opex, insurance and connection costs. The other data required: the capacity of the plant (5MW) and a life time of the plant (20 years). The data is given below taken from (IEA-IRENA, 2015), which also shows that h-WTES can save 12% on construction and 50% on fixed opex.

Electrical Wind Turbine	
Pre-development period	4 years
Pre-development costs	110 £/kW
Construction Period	2 years
Construction Costs	1200 £/kw
Infrastructure Cost	£ 3300
Fixed Opex	23000£/MW/yr
Variable Opex	5 £/MWh
insurance	1400 £/MW/yr
Connection	3100 £/MW/yr
Energy storage	11 £/MWh-th

Heat Wind Turbine WTES	
Pre-development period	4 years
Pre-development costs	110 £/kW
Construction Period	2 years
Construction Costs	840 £/kw
Infrastructure Cost	£ 3300
Fixed Opex	11600£/MW/yr
Variable Opex	2.5 £/MWh
insurance	1400 £/MW/yr
Connection	3100 £/MW/yr
Energy storage	11 £/MWh-th

Table 2: Capital costs and Operating & Variable costs from literature. (NICOLÁS NITTO, 2017)

$$\text{NPV of Total Costs} = \sum_{n} \frac{\text{total capex and opex costs}_n}{(1 + \text{discount rate})^n} \tag{1}$$

$$NPV \text{ of Energy Generation} = \sum_{n} \frac{\text{net generation}_n}{(1 + \text{discount rate})^n} \tag{2}$$

$$\text{Levelised Cost of Energy} = \frac{\text{NPV of Total Costs}}{\text{NPV of Energy Generation}} \tag{3}$$

Where n = period (Lifetime of the plant)

2.2 Results and Discussion

This current section shows the results obtain from the economic assessment of the wind heat & electrical turbine with thermal energy storage. It is composed of two parts: firstly, the LCOE results from 2017 data and second part shows the LCOE for 2020 and 2025.

General Results – LCOE Estimation

Parameters	Units	e-wind turbine	h-wind turbine	Traditional heating source (natural gas)
Capacity	kW	5000	5000	
Pre-Development Costs	£/MWh	4.6	4.6	
Construction costs	£/MWh	43.7	37.7	
Energy Storage	£/MWh	11.4	11.4	
Fixed O&M (+insurance+ connection)	£/MWh	9.8	9.4	
Variable O&M	£/MWh	5.0	4.5	
Electrical driven HP	£/MWh	1.64		
LCOE	£/MWh	65.0	56.2	18.7

Table 3: This shows the net present value calculated using the above equation for e-wind turbine and h-wind turbine with thermal energy storage.

LCOE results for 2020 and 2025

h-wind turbine

	Units	2017	2020	2025	
Est. h-wind turbine	£/MWh-t	56.2	52.2	50.5	15% less than onshore wind turbine
Thermal Energy Storage	£/MWh-t	11.4	10.3	9.2	SunShot target: £10.3/MWh-t @2020 10% reduction after 5 years.
COE of h-WTES	£/MWh	67.6	62.5	59.7	

Table 4: shows the LCOE costs for h-wind turbine with thermal energy storage

e-wind turbine

	Units	2017	2020	2025	
Onshore Wind Turbine	£/MWh-t	63.36	61.4	59.44	Data from BEIS, Nov 2016
Electrical Heater	£/MWh	1.64	1.60	1.56	Data from IRENA 2016
Efficiency of heater	%	100%	100%	100%	Data from IRENA 2016
Thermal Energy Storage	£/MWh-t	11.4	10.3	9.2	SunShot target: £10.3/MWh-t @2020 10% reduction after 5 years.
COE of h-WTES	£/MWh	76.41	73.3	70.2	

Table 5: shows the LCOE costs for e-wind turbine with thermal energy storage

<u>LCOE for h-Wind Turbine with Thermal Energy Storage</u>

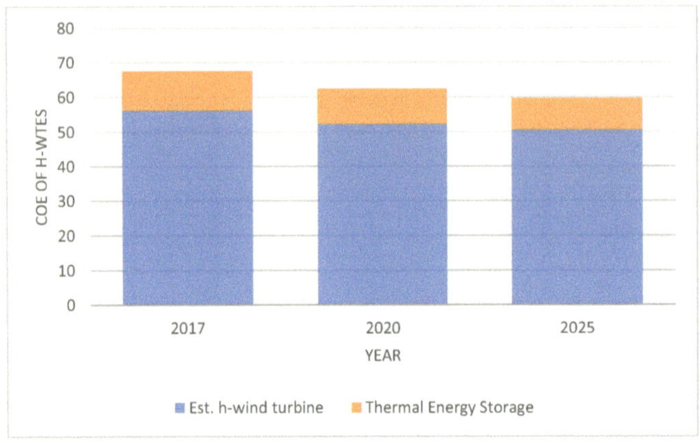

Figure 7: This illustrates the LCOE for h-wind turbine with storage for 2020 and 2025.

<u>LCOE for e-Wind Turbine with thermal energy storage</u>

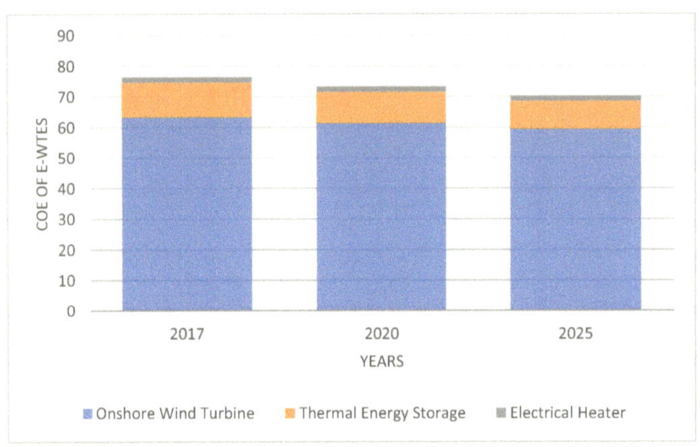

Figure 8: This illustrates the LCOE for e-wind turbine with storage for 2020 and 2025.

2.21 Estimated energy cost

The estimated energy cost for the electrical-wind turbine (e-WTES) with storage & electrical heater and heat-wind turbine (h-WTES) with storage is shown in Fig 7 and 8. The result indicates that the

LCOE$_{HEAT}$ of direct heat production h-WTES is 12% lower than indirect heat production e-WTES, with exact same thermal energy cost. The reason for this is that e-WTES employ electrical driven heater (100% efficiency) to convert the electricity into heat which increases the cost of e-WTES and further, according to (IEA-ETSAP and IRENA, 2015) h-WTES can save around 12% compared to a traditional wind turbine on construction which decreases the cost of h-WTES. Therefore, it can be concluded that it is more economically feasible to generate heat from h-WTES with thermal storage for industrial application than e-WTES with thermal storage.

According to (Okazaki, Shirai and Nakamura, 2015), the cost of thermal energy storage is 60-70% cheaper compared to other storage application such as (Backup thermal and Battery) and is favourable for WTES since wind repetition time is longer than 24hr and requires larger storage capacity.

Additionally, table 3 shows the levelized cost of heat range for industrial boiler, which is commonly used to produce heat in the industries, the LCOE$_{HEAT}$ ranging between 15-20 £/MWh (ec.europa.eu, 2016). As expected and according to the literature reviewed, the renewable technologies with thermal storage present the most expensive LCOE for heat production purposes compare to the industrial boiler. Further, the graph shows there is a decrease in the thermal energy cost, 10% reduction every 5years, which is also forecast by (SunShot target 2016), this could be due to further developments in thermal energy storage. Also, the graph shows that there is an overall decrease in the LCOE of h-WTES 12% and e-WTES 8% by 2025. Moreover, it can be assumed that the technologies such WTES can meet the LCOE of the traditional gas boiler in future, mainly due to the current energy crisis, CO_2 tax, Carbon capture and reduction in fossil fuel.

3.0 An effectiveness-NTU technique

An effectiveness-NTU technique for characterising PCM thermal storage system for WTES application

This part of the research includes the design of thermal storage system with the phase change material using mathematical formulation. A simplified method which was proposed by (Tay, Belusko and Bruno, 2012) is being used to characterise the phase change thermal storage system based on 1D phase change. An analytical method developed using the effectiveness-number of transfer units (ε-NTU) technique for a tank filled with PCM, with HTF flowing through tubes inside the shell as shown in Fig 9. (Tay, Belusko and Bruno, 2012)

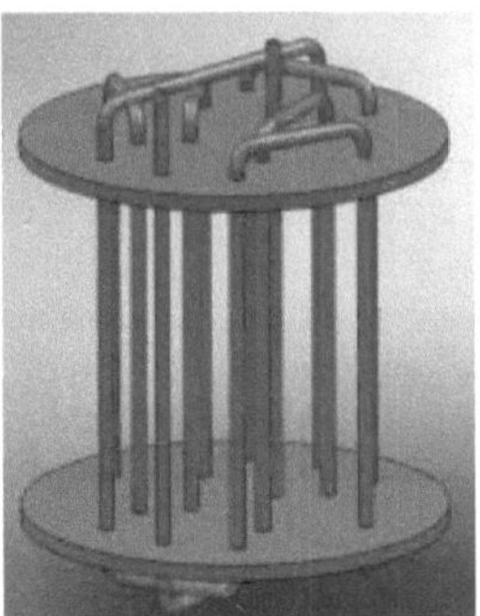

Figure 9: Schematic of one tube tank

Previous research by (Chan and Tan, 2006) (Tan, 2008) have empirically demonstrated that effectiveness-NTU can be used for the PCM thermal storage system to characterise the heat transfer and the result showed that the heat exchange effectiveness clearly correlates with the number of transfer units (NTUs). The CFD model of the PCM within a tube in tank developed by (Tay et al., 2014) also validate the result of effectiveness-NTU. From the experimental result of (Tay, Belusko and Bruno, 2012), it was found that the tube-in-tank with sufficient heat transfer area has the highest energy storage density with capacity factor above 90%, identified by the average heat exchange effectiveness which defines the performance of the storage system.

3.1 Description of the Latent-Thermal Energy Storage system

The system consists of a metal shell and a 15 x 15 matrix of copper pipes (see figure below Fig 10). The heat transfer fluid HTF in this work is mineral oil as it provides reliable heat transfer and the properties are presented in table 7. The HTF flows through the tubes exchanging heat with the PCM filled inside the shell (Fig 11). The dimensions of HTF tube is (1mm thick and 12.5mm i.d.), and 4.0m long. NaNO$_3$-NaOH has been used as a PCM (Pereira da Cunha and Eames, 2016), its properties are presented in table 6.

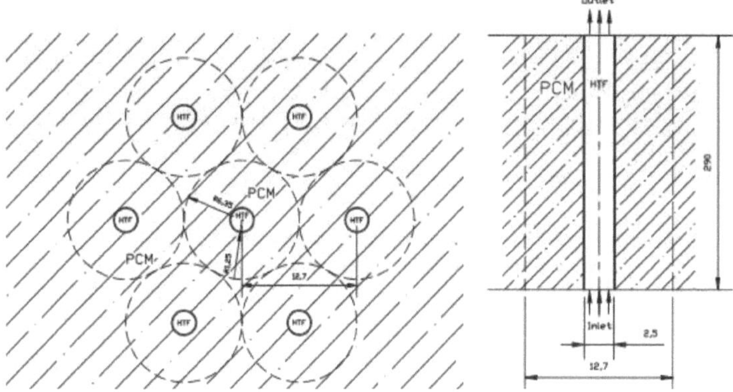

Figure 10: a Simple schematic unit of PCM-TES system Right) Single heat transfer tube and left) unit cross-section and all the dimensions are present in cm. (Colella, Sciacovelli and Verda, 2012)

The Phase Change Material PCM

The selected PCM was identified from the Phase Change Material Review (Pereira da Cunha and Eames, 2016) and the properties of the material are presented in Table 6. The selected PCM was based on being representative of typical eutectic mixture and salt hydrates, with a melting point suitable for Food and Paper industrial sector. According to the review, salt hydrates are promising for applications below 100^0C and eutectic mixtures from 100 to 300^0C. Appropriate selection of the phase change material is beyond the scope of this research.

Organic compounds were not selected as a PCM material as they have very low thermal conductivity (range from 0.1 to 0.7 W/mK), they usually require a mechanism to enhance their heat transfer rate.

Eutectic Compound NaNO$_3$-NaOH Properties	
T_{PCM}	250^0C
C_{pPCML}	1.86kJ/kg.K
H_L	160kJ/kg
P_{PCM}	2241kg/m^3
K_{PCMS}	0.60W/mK
K_{PCML}	0.66W/mK

Table 6: PCM properties

Heat Transfer Fluid Properties

Mineral-oil has been chosen as a Heat transfer fluid as it provides reliable heat transfer and has the performance benefits of low fouling, high thermal stability, practically non-toxic and environmentally friendly.

Mineral Oil Properties	
C_{pHTFL}	0.162kJ/kg.K
$U_{viscosity}$	0.00987 kg/ms
K_{HTF}	0.162W/mK
p_{HTF}	850 kg/m^3
Boiling point	358^0C

Table 7: HTF properties (Therminol.com, 2017)

4.2 Mathematical Model

The formulation of the heat flow is a one dimensional between the PCM and the HTF at the phase change profile. It is based on the internal temperature measurement of PCM by (Tay, Belusko and Bruno, 2012) where it was found that the phase change was uniform throughout the PCM tank. The NTU is calculated from the thermal resistance to heat flow within the tube wall, HTF and the part of PCM which undergoes a phase change.

<u>Assumptions</u>

- The phase change occurs in a cylindrical pattern and one-dimensional Fig 11.
- Constant inlet temperature and velocity of the HTF and the outer wall of the thermal storage system remaining adiabatic.
- System initial temperature is uniform.
- Natural convection has been ignored.
- The thermos physical properties of the PCM, the tube wall and the HTF are constant.
- Single length tube arrangement. Bend impact is ignored.
- S_{round} – represents the conduction between two concentric cylinders so in this case the heat flow from the tube to the phase change front Eq 5.
- R_{max} – The phase change front varies with time, start at tube wall and increase to max radius R_{max}.

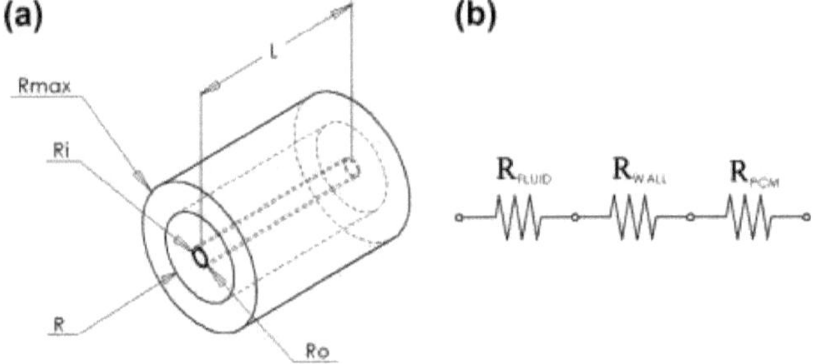

Figure 11: a) simplified model of the storage unit. b) thermal circuit

The NTU can be presented by the following equation at any point in time.

$$NTU = \frac{UA}{mC_p} = \frac{1}{R_T . m . C_p} \qquad (1)$$

To formulate the storage effectiveness, the thermal resistance needs to be determined which is a function of the overall heat transfer coefficient between the PCM and HTF, and the heat transfer area. The overall thermal resistance R_T is given by the equation below (2) and (3), where R_{PCM} is the resistance in the PCM, R_{WALL} resistance in the tube wall and R_{FLUID} resistance of the heat transfer fluid defined by forced internal convection. (Tay, Belusko and Bruno, 2012)

$$R_T = R_{HTF} + R_{WALL} + R_{PCM} \qquad (2)$$

$$R_T = \frac{1}{(2\pi R_I L h_f)} + \frac{\ln\left(\frac{R_O}{R_I}\right)}{(2\pi k_w L)} + \frac{1}{S_{round} k_{PCM}} \tag{3}$$

$$S_{round} = \frac{(2\pi L)}{\ln\left(\frac{R}{R_O}\right)} \tag{4}$$

$$R_T = \frac{1}{(2\pi R_I L h_f)} + \frac{\ln\left(\frac{R_O}{R_I}\right)}{(2\pi k_w L)} + \frac{\ln\left(\frac{R}{R_O}\right)}{2\pi k_{PCM} L} \tag{5}$$

The shape factor varies with time; this can be calculated based on phase change fraction δ for PCM, which is the amount of material that has yet to change phase. For tube surrounded by PCM (cylindrical volume), the phase change fraction δ is define by the following equation (6). Where, A_o outer tube area, A_{MAX} maximum area after phase change and A_r is the changing area of PCM. (Tay, Belusko and Bruno, 2012)

$$\delta = \frac{(A_r - A_o)}{(A_{max} - A_o)} = \frac{(R^2 - R_0^2)}{(R_{max}^2 - R_o^2)} \tag{6}$$

$$R = \left\{ \delta \left(R_{max}^2 - R_o^2 \right) + R_o^2 \right\}^{\frac{1}{2}} \tag{7}$$

The overall thermal resistance R_T is defined by the following equation. Substituting Equation (7) into (5).

$$R_T = \frac{1}{(2\pi R_I L h_f)} + \frac{\ln\left(\frac{R_O}{R_I}\right)}{(2\pi k_w L)} + \frac{\ln\left(\frac{\left\{ \delta \left(R_{max}^2 - R_o^2 \right) + R_o^2 \right\}^{\frac{1}{2}}}{R_O}\right)}{2\pi k_{PCM} L} \tag{8}$$

Therefore, the effectiveness can be defined by the following equation (9) in terms of R_T, at certain phase change fraction and mass flow rate.

$$\varepsilon = 1 - \exp(-1/mC_p R_T) \tag{9}$$

Set of equations below are used for the calculation of h_f.

$$\text{Prandtl number, } Pr = \mu_f C_p / k_f \tag{10}$$

$$\text{Reynolds number, } Re = md_i / A_c \mu_f$$

Nusselt's number,

$$\text{Laminar flow, } Nu = \frac{3.66 + (0.0668(d_i/L)R_eP_r)}{(1 + 0.04[(d_i/L)R_eP_r]^{\frac{2}{3}}} \tag{11}$$

Turbulent flow, $Nu = 0.023\,Re^{0.8}\,Pr^{n}$ where: n = 0.4 for heating and 0.3 for cooling

$$h_f = (Nuk_f)/d_i \tag{12}$$

Where: A_c is the cross-section area of the inner tube, di inner diameter, m is the HTF mass flow rate, C_p is the HTF specific heat capacity and μ_f is the dynamic viscosity.

The heat exchange effectiveness changes with the phase change process, therefore it is important to determine the average effectiveness to obtain the performance of the system. The average effectiveness is defined as follows.

$$\bar{\varepsilon} = \int_0^1 (1 - \exp(-1/mC_pR_T))d\delta \tag{13}$$

4.3 Results and Discussion

The ε-NTU technique has been used for characterising a latent/PCM-TES system integrated into WTES for industrial application such as Food and Paper sector, which requires heating of a fluid stream (hot water, hot air streams) and heating of some reservoir for various applications. MATLAB was used to perform all the calculation, and the results are presented in the graphs below.

In practice, a PCM thermal storage system can be described as a heat exchanger in which the heat transfer fluid HTF exchanges heat with the PCM at the phase change temperature (Kalaiselvam and Parameshwaran, 2016). The performance of a PCM thermal storage device can be described by the heat exchange effectiveness. The effectiveness is defined as the ratio of the actual heat transfer to the maximum possible heat transfer, which is defined by the maximum temperature difference between the heat transfer fluids. A PCM thermal storage system can also be characterised by its effectiveness Equation (9), (Ismail, Quispe and Henríquez, 1999) (Belusko, Halawa and Bruno, 2012).

The maximum effectiveness of the system typically occurs when the HTF outlet temperature equal to the PCM temperature as this is illustrated in Fig 12 and Fig 13 where at mass flow rate of 0.001kg/s, the effectiveness of the system is around 95% with the outlet temperature of HTF is 245⁰C which equals to the PCM temperature. The effectiveness is a very useful design parameter, for both discharging and charging processes. In a thermal storage system with PCM, the exit temperature of HTF during the discharging process should meet the system specifications, (In this case, the average temperature of the food and the paper industrial sector should be around 170⁰C (IEA-ETSAP and IRENA, 2015)). Therefore, the minimum required effectiveness after which the thermal storage system is ineffective is a performance parameter of the system design (Iqbal, 2014) (Kreith and Bohn, 2001). However, for both discharging and charging processes different parameters can be changed such as mass flow rate, mass flux and length of the tube to maximise the effectiveness, minimise the exergy losses and maximise the useful energy that is stored. The Fig 12, clearly illustrates that the average effectiveness of the system is the function of mass flow rate of the HTF. As the flow rate of the HTF decreases the average effectiveness of the thermal storage system increases so does the average outlet temperature of the HTF (Fig 13). Therefore, the above parameter has been optimised to satisfy the minimum outlet temperature and the average effectiveness of the industrial processes (effectiveness = 31% and outlet HTF temperature = 175⁰C) (Belusko, Halawa and Bruno, 2012)

The analysis of the average effectiveness of thermal energy storage only considers the latent energy stored in the PCM and ignores the sensible energy stored, the reason for ignoring the sensible energy is because most of the energy is stored in the latent phase for the PCM storage system and the maximum exergy efficiency typically occurs when the melting point of the phase change material is the average of heat sink and heat source temperature. Therefore, the system is only applicable to the low-temperature difference.

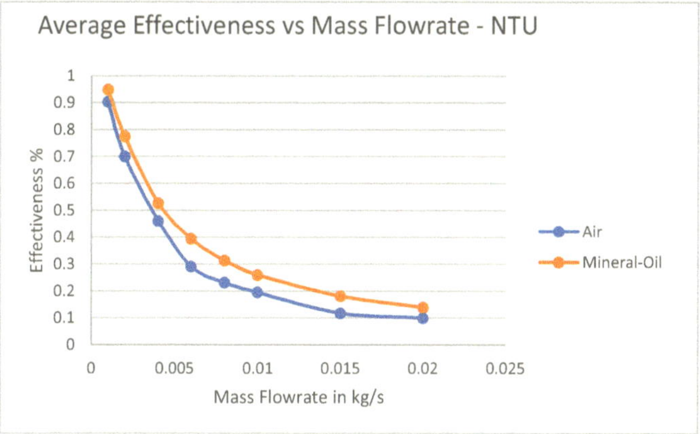

Figure 12: Graph shows the average effectiveness vs. mass flowrate of two different HTF Air and mineral-oil for Latent/PCM-Thermal Energy Storage Device. Obtained using the ε-NTU technique as shown above.

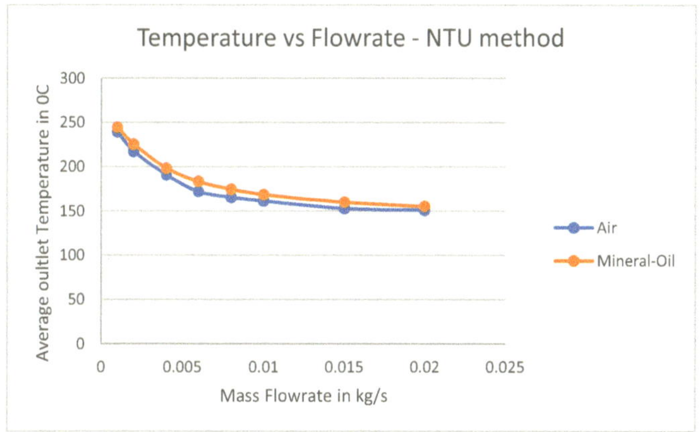

Figure 13: Graph shows the average outlet temperature ⁰C vs mass flowrate of two different HTF air and mineral-oil for Latent/PCM-Thermal Energy Storage Device. Obtained using the ε-NTU technique as shown above.

Figure 12, shows the calculated average effectiveness for the melting process using Eutectic Compound $NaNO_3$-NaOH as a phase change material with the melting point of 250⁰C, using two different heat transfer fluid, Air (blue line) and Mineral Oil (orange line), overall showing good agreement with an error of around 14%. A sudden change in Blue curve occurs at ṁ = 0.006kg/s. This

is because of the change of condition of Air from the laminar regime to turbulent regime. At the point
ṁ = 0.006kg/s, the Reynolds number reached > 2000 values of turbulent flow, making the Nusselt to
increase and causing the overall thermal resistance to decrease. Therefore, water is not suitable to be
used as an HTF for this process.

The result from the ε-NTU technique also shows good agreement with the experimental result,
conducted by (Lik ab., 2017) for Solar Tower Power Plants using the same PCM and HTF. The result
has an average absolute error of around 10%; this is mainly due to the natural convection which occurs
during the melting process of the experiment. The report by (Bédécarrats et al., 1996) also shows that
the effectiveness-NTU technique does not take into account the natural convection which depends on
the system temperature difference but, the effectiveness-NTU method is intended to be temperature
independent therefore no natural convection. (The graph of experimental result and modelling result
present in appendix 1).

Parameters	Results
Mass flow rate of HTF (kg/s)	0.02
Mass flux ṁ/A (kg/sm^3)	0.0003
Number of transfer units (NTU)	0.55
Total thermal resistance K/W	0.134
Average effectiveness	0.313
Overall heat transfer rate kW	115
Average HTF outlet temperature ^{0}C	175

Table 8: Results obtained from the effectiveness-NTU technique

By analysing the existing literature and the results of the model, it can be concluded that the technique
ε-NTU proposed by (Tay, Belusko and Bruno, 2012), can approximately (since it only consider the
average temperature/effectiveness of the system) predict the average effectiveness of the system
(which describes the performance of the device) with a high heat transfer rate for discharging and
charging processes for PCM-Storage system (115kW) Table 8. Therefore, this design can effectively be
integrated into WTES to provide a large amount of heat demand for industrial applications. Further,
the table 4, below shows some of the important parameter obtained from the ε-NTU mathematical
model which includes the energy capacity (total amount of energy that can be stored), Maximum
charge/discharge power, energy storage duration and round-to-round efficiency.

Parameters	Results
Overall heat transfer Charge/Discharge rate kW	115
Energy storage capacity MWh for the unit	1.085
Total energy Storage capacity GWh/yr	1.41
Energy storage duration hours	9.4
Capacity Factor (UK average for the wind)	0.22
WTES system capacity	440kW
Carbon emission reduction metric tons per day	18.3

Table 9: further Results obtained from the effectiveness-NTU technique

Wind Renewable system such as Wind thermal energy system (WTES) are intermittent and cannot be
used as the baseload energy source as there inevitably exists a time discrepancy between demand
and generation. Therefore, thermal energy storage (TES) has become a crucial technology for energy
crisis and, plays an important role for renewable applications as it makes the operation more efficient.
The above result shows that PCM-Thermal Storage system integrated with WTES can store large

amount of surplus latent energy at times when the industrial heat demand is low and can provide stable heat at peak times for longer period (9.4 hours) with discharge power of 115kW and average outlet temperature of 175⁰C, which is enough for industrial sectors such as Food, Beverages and Paper Table 10. This means that by having TES integrated into the system can provide a significant amount of energy for a longer period as illustrated in Fig 14.

Industrial Sector	Temperature Range ^{0}C	Heat Intensiveness
Food	30 - 120	Low
Beverages	60 - 90	Low
Paper	60 - 150	Low

Table 10: shows the temperature range of industrial sector

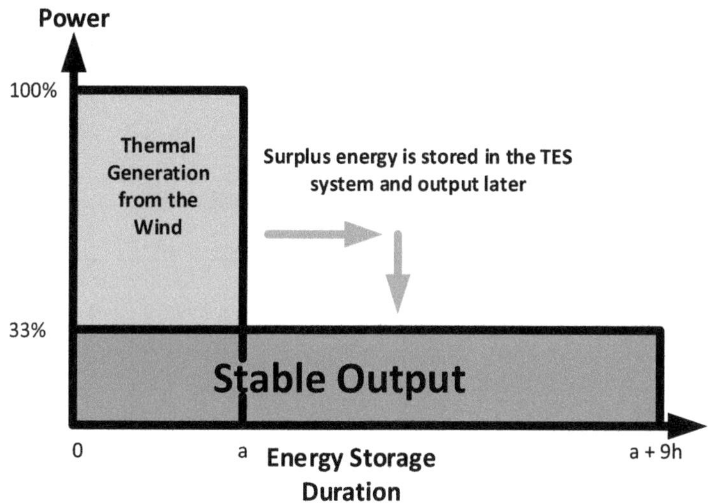

Figure 14: Shows the Stable output of the thermal energy (PCM)

Latent/PCM – Thermal energy storage system has the total storage energy capacity of 1.41GWh per year with the average effectiveness of 31% to achieve the specified temperature. 2MW-WTES with PCM-TES has a system capacity of 440kW (considering the UKs wind capacity factor 0.22 (Gov.uk, 2017)) with the storage discharge rate of 115kW. According to (Law, Harvey and Reay, 2017) research report, food industries in the UK consumes around 28.767GWh of energy per day, which is around 2 – 2.8GWh/yr for a single plant in the UK. Therefore, a PCM-Storage system with WTES has enough energy capacity (440kW) to supply heat demand for the food plant for application such as drying, washing, pasteurising, boiling, bleaching, sterilising and heat treatment (IEA-ETSAP and IRENA, 2015).

Sustainability is increasing concern in the food industry, and the various report shows that the food industry is lagging in the environmental performance compared to other sectors. This is mainly because the food/drink industry plays a significant role in environmental impact as they currently account for 7 million tonnes of carbon emission each year (Gov.uk, 2017). Interestingly, by having a 2MW WTES with a PCM-storage system for the food industry, can reduce 625 tons of CO_2 emission per year from burning fuel such as natural gas (the value of carbon saved is calculated using the current DECC guidelines). It should be noted that by having this system in the different sector can help the UK to meet their emission target. (Gov.UK, 2017)

Comparison of PCM storage vs. Sensible storage tanks

Sensible heat storage specifications

Heat Transfer Fluid	Mineral Oil
Inlet temperature HOT	310^0C
Outlet temperature Cold	$40°C$

Table 11: Sensible heat storage specifications

As expected, the calculated result shows that the sensible heat storage system requires larger storage unit (5 x bigger) than PCM – Storage Unit to store the same amount of energy (1.085MWh). The reason being, the PCM has very high energy storage densities at a constant temperature (Liu et al., 2016) and are an ideal candidate for heat storage. An experiment conducted by (McKeever, 2015) shows that the PCM-TES can hold 6.5 times more heat than the SH-TES which allowed the CHP to run longer at a peak period and also eliminated the use of backup gas boilers.

Result

	Sensible Heat – Storage System	**PCM – Storage system**
Volume m^3	52.2	11.4

Figure 15: Comparison between the PCM storage vs. the Sensible Storage tank

More thermal properties of the commonly used liquid material for sensible thermal storage are present in the appendix.

5.0 Conclusion

Fossil fuel plays an important role in the society, and they are commonly used for the heat and power production for the commercial and residential applications. According to UKERC Research Report, heat accounts for around 47% of the total energy consumption in the UK and 33% of carbon emissions.

Approximately, 80% of this heat account for space and water heating. At the same time, a large amount of heat is being used by the industry sector 24%, for industrial processes. The corporation of thermal energy storage system is very beneficial for CO_2 emission reduction and energy saving. And, it is the key technology to balancing the energy system stable with the increasing capacity from wind.

Limited research on the concept of Wind-Powered Thermal Energy System (WTES) recently proposed by (2015, Okazaki et al.) with Latent-Thermal Energy Storage to produce heat for industrial application and also, the economics of WTES and electrical-WTES with thermal energy storage for heat generation is unknown.

A study of the potential and probable costs of different WTES with thermal energy storage and the designing of the latent-TES using effectiveness-NTU modelling suitable for industrial sector was conducted.

It was identified that it is more economically feasible to generate heat from h-WTES (Direct heat production) with thermal storage than e-WTES (indirect heat production). The reason for this was that e-WTES employed electrical driven heater to convert the electricity into heat which increased the $LCOE_{thermal}$ of e-WTES to 63.36 £/MWh-t and further, h-WTES saved around 30% on construction which decreased the LCOE of h-WTES to 56.2 £/MWh-t. Additionally, the $LCOE_{thermal}$ was compared with the traditional industrial gas boiler ($LCOE_{thermal}$ = 15-20 £/MWh), the results suggested that the renewable technologies with TES present the most expensive LCOE for heat production compare to the industrial gas boiler. However, these technologies can meet the LCOE of the traditional gas boiler in future, mainly due to the current energy crisis, CO_2 tax, Carbon capture and reduction in fossil fuel.

For the second part of the research latent-TES (present higher energy storage density) was characterised using the effectiveness-NTU technique based on 1D phase change. It was investigated that the technique effectiveness-NTU can accurately predict the average effectiveness of the system (performance) with high heat transfer rate for discharging and charging processes (115kW). Therefore, the design can effectively be integrated into WTES to provide a significant amount of heat demand for industrial plants such as food and paper. It was also, found that latent-TES can hold 6.5 times more heat than sensible-TES. Finally, it was identified that by having a 2MW WTES with latent-TES for the food industry, can reduce 625 tons of CO_2 emission per year from burning fuel such as natural gas. It should be noted that by having this system in different sector can help UK to meet their emission target.

Further research is needed to examine the impact of combining WTES with other thermal plants such as geothermal, CSP, biomass with different thermal energy storage material or technologies (as it can be shared) to generate heat for industrial applications.

6.0 References

Bédécarrats, J., Strub, F., Falcon, B. and Dumas, J. (1996). Phase-change thermal energy storage using spherical capsules: performance of a test plant. *International Journal of Refrigeration*, 19(3), pp.187-196.

Belusko, M., Halawa, E. and Bruno, F. (2012). Characterising PCM thermal storage systems using the effectiveness-NTU approach. *International Journal of Heat and Mass Transfer*, 55(13-14), pp.3359-3365.

Belusko, M., Tay, N., Liu, M. and Bruno, F. (2016). Effective tube-in-tank PCM thermal storage for CSP applications, Part 1: Impact of tube configuration on discharging effectiveness. *Solar Energy*, 139, pp.733-743.

Chakirov, R. and Vagapov, Y. (2011). Direct Conversion of Wind Energy into Heat Using Joule Machine. *2011 International Conference on Environmental and Computer Science*, (Bonn-Rhein-Sieg University of Applied Science).

Chan, C. and Tan, F. (2006). Solidification inside a sphere—an experimental study. *International Communications in Heat and Mass Transfer*, 33(3), pp.335-341.

Colella, F., Sciacovelli, A. and Verda, V. (2012). Numerical analysis of a medium scale latent energy storage unit for district heating systems. *Energy*, 45(1), pp.397-406.

ec.europa.eu (2016). *Subsidies and costs of EU energy Final report*.

gov.uk (2016). *ELECTRICITY GENERATION COSTS*. Department of business, energy and industrial strategy.

Gov.uk (2017). *Department of Energy & Climate Change*.

Gov.uk. (2017). *Digest of UK Energy Statistics (DUKES) - GOV.UK.* [online] Available at: https://www.gov.uk/government/collections/digest-of-uk-energy-statistics-dukes [Accessed 2 Aug. 2017].

Hedegaard, K. and Münster, M. (2013). Influence of individual heat pumps on wind power integration – Energy system investments and operation. *Energy Conversion and Management*, 75, pp.673-684.

IEA-ETSAP and IRENA (2015). *Solar Heat for Industrial Processes*. Technology Brief.

Iqbal, M. (2014). *Thermal effectiveness of a split-flow heat exchanger.*. pp.20-86.

IRENA (2012). *costs analysis series concentrating solar power. International Renewable Energy Agency, pp.1-2.*. Renewable energy technologies.

Ismail, K., Quispe, O. and Henríquez, J. (1999). A numerical and experimental study on a parallel plate ice bank. *Applied Thermal Engineering*, 19(2), pp.163-193.

Kalaiselvam, S. and Parameshwaran, R. (2016). *Thermal energy storage technologies for sustainability*. p.60.

Kreith, F. and Bohn, M. (2001). *Principles of heat transfer*. Australia: Brooks/Cole Pub.

Law, R., Harvey, A. and Reay, D. (2017). Opportunities for Low-Grade Heat Recovery in the UK Food Processing Industry. *Sustainable Thermal Energy Management in the Process Industries International Conference (SusTEM2011)*, (Newcastle University).

Lee; Jean L (2012). *ON-DEMAND GENERATION OF ELECTRICITY FROM STORED WIND ENERGY*. 20120326445.

Li, G. and Zheng, X. (2016). Thermal energy storage system integration forms for a sustainable future. *Renewable and Sustainable Energy Reviews*, 62, pp.736-757.

Lik ab (2017). *Validation of effectiveness-NTU*. Master thesis (Pakistan University): Energy.

Liu, M., Steven Tay, N., Bell, S., Belusko, M., Jacob, R., Will, G., Saman, W. and Bruno, F. (2016). Review on concentrating solar power plants and new developments in high temperature thermal energy storage technologies. *Renewable and Sustainable Energy Reviews*, 53, pp.1411-1432.

Liu, M., Tay, N., Belusko, M. and Bruno, F. (2015). Investigation of Cascaded Shell and Tube Latent Heat Storage Systems for Solar Tower Power Plants. *Energy Procedia*, 69, pp.913-924.

Matsuo, T. and Okazaki, T. (2017). A Basic Theory of Induction Heating for a Wind-powered Thermal Energy System. *IEEE Transactions on Magnetics*, pp.1-1.

McKeever, M. (2015). Validating the Performance of a Prototype Phase Change Material for a Thermal Energy Storage Tank, Connected to a Micro-CHP. *SDAR* Journal of Sustainable Design & Applied Research*, 3(1).

Meibom, P., Kiviluoma, J., Barth, R., Brand, H., Weber, C. and Larsen, H. (2007). Value of electric heat boilers and heat pumps for wind power integration. *Wind Energy*, 10(4), pp.321-337.

NICOLÁS NITTO, A. (2017). WIND POWERED THERMAL ENERGY SYSTEMS (WTES) A TECHNO-ECONOMIC ASSESSMENT OF DIFFERENT CONFIGURATIONS. *UNIVERSITY OF OLDENBURG*.

Okazaki, T., Shirai, Y. and Nakamura, T. (2015). Concept study of wind power utilizing direct thermal energy conversion and thermal energy storage. *Renewable Energy*, 83, pp.332-338.

Pereira da Cunha, J. and Eames, P. (2016). Thermal energy storage for low and medium temperature applications using phase change materials – A review. *Applied Energy*, 177, pp.227-238.

Radcliffe, J. (2015). Thermal energy storage in Scotland. *University of Birmingham*.

Research Report (UKERC) (2014). *The Future Role of Thermal Energy Storage in the UK Energy System: An assessment of the Technical Feasibility and Factors Influencing Adoption.*.

Tan, F. (2008). Constrained and unconstrained melting inside a sphere. *International Communications in Heat and Mass Transfer*, 35(4), pp.466-475.

Tay, N., Belusko, M. and Bruno, F. (2012). An effectiveness-NTU technique for characterising tube-in-tank phase change thermal energy storage systems. *Applied Energy*, 91(1), pp.309-319.

Tay, N., Belusko, M. and Bruno, F. (2012). Experimental investigation of tubes in a phase change thermal energy storage system. *Applied Energy*, 90(1), pp.288-297.

Tay, N., Belusko, M., Castell, A., Cabeza, L. and Bruno, F. (2014). An effectiveness-NTU technique for characterising a finned tubes PCM system using a CFD model. *Applied Energy*, 131, pp.377-385.

Therminol.com. (2017). *Therminol® XP | Therminol*. [online] Available at: https://www.therminol.com/products/Therminol-XP [Accessed 12 Aug. 2017].

Appendices

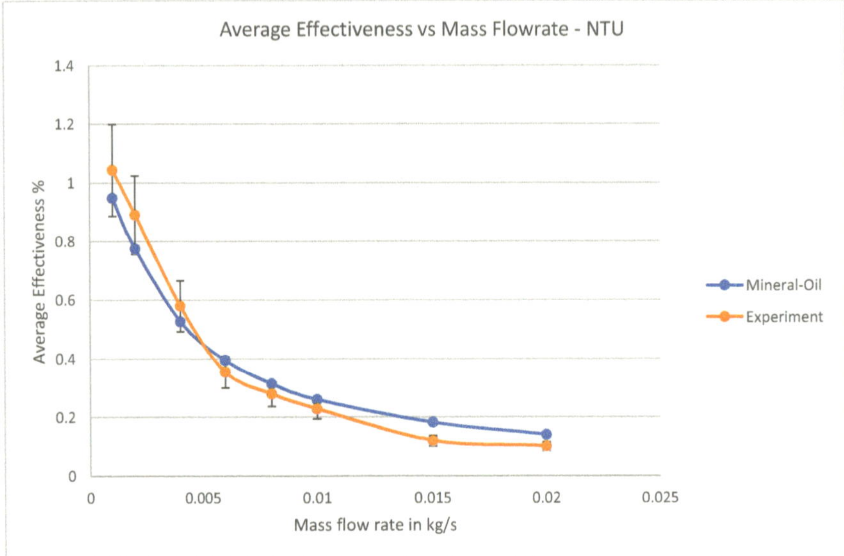

Appendix 1: The result from the ε-NTU technique also shows good agreement with the experimental result, conducted by (Lik ab., 2017) for Solar Tower Power Plants using the same PCM and HTF. The result has an average absolute error of around 10%.

Appendix 2: Thermal Properties of Sensible Thermal Storage materials

Thermal properties of commonly used liquid materials for sensible thermal storage:

| Storage medium | Temperature | | Average density (kg/m³) | Average thermal conductivity (W/m K) | Average heat capacity (kJ/kg K) | Volume specific heat capacity (kWh/m³) | Media costs/kg (US$/kWh) | Media costs/kWh (US$/kWh) |
	Cold (°C)	Hot (°C)						
Mineral oil	-10	300	770	0.12	2.6	55	0.30	4.2
Synthetic oil	13	350	900	0.11	2.3	57	3.00	43.0
Silicone oil	-40	400	900	0.10	2.1	52	5.00	80.0
Nitrite salts	250	450	1825	0.57	1.5	152	1.00	12.0
Nitrate salts	265	565	1870	0.52	1.6	250	0.50	3.7
Carbonate salts	450	850	2100	2.0	1.8	430	2.40	11.0
Liquid sodium	270	530	850	71.0	1.3	80	2.00	21.0

Appendix 3: summary of TES with various application

Thermal Energy Storage materials

		Sensible heat	Phase change	Thermo-chemical
	Example materials	Water, gravel, pebble, soil …	Organics, inorganics …	Metal chlorides / hydrides / oxides …
	Type of TES system	Water-based (water tank, aquifer), rock or ground-based	Packaged in passive storage systems	Adsorption/absorption, chemical reaction, at high temperatures
Technology applications of TES materials	Seasonal heat	Demonstrated		
	CHP	Commercial		
	Cooling/chilling	Commercial	Deployed	
	Domestic TES	Commercial	In Development	In Research
	Building fabric	Deployed	Demonstrated	
	Cryogenic	Demonstrated	Demonstrated	
	Solar thermal	In early deployment	In early deployment	
	Pumped thermal	In development		
	Advantages	Environmental, cost, simplicity, reliability	Higher energy density, constant temperature	Highest energy density, low loss for long duration storage
	Disadvantages	Low energy density, heat losses, site construction capex	Lack of stability, corrosion, high cost of storage materials	Poor heat transfer, cyclability, high cost of storage material

Technology application features

Use/location	Timescale	Energy scale
Balancing summer/ winter heat demand	Months	GWh
Balancing electrical/heat network demand	Hours - days	10s MWh
Reduction in daily peak; commercial properties	Hours	10s MWh
Reduction in daily peak; households	Hours	10s kWh
Reduction in daily peak; Domestic/commercial	Hours	kWh and above
Grid integration of variable renewables	Hours – days	10s –100s MWh
Smoothing output from solar thermal	Hours	10s –100s MWh
Grid integration of variable renewables	Hours	Up to 10s MWh

This section shows the integration of thermal energy storage into different applications.